GROSS AND DISGUSTING THINGS



# GROSS AND DISGUSTING STUFF IN YOUR BODY

A Crabtree Branches Book

Julie K. Lundgren

Crabtree Publishing
crabtreebooks.com

# School-to-Home Support for Caregivers and Teachers

This high-interest book is designed to motivate striving students with engaging topics while building fluency, vocabulary, and an interest in reading. Here are a few questions and activities to help the reader build upon his or her comprehension skills.

## Before Reading:

- *What do I think this book is about?*
- *What do I know about this topic?*
- *What do I want to learn about this topic?*
- *Why am I reading this book?*

## During Reading:

- *I wonder why...*
- *I'm curious to know...*
- *How is this like something I already know?*
- *What have I learned so far?*

## After Reading:

- *What was the author trying to teach me?*
- *What are some details?*
- *How did the photographs and captions help me understand more?*
- *Read the book again and look for the vocabulary words.*
- *What questions do I still have?*

## Extension Activities:

- *What was your favorite part of the book? Write a paragraph on it.*
- *Draw a picture of your favorite thing you learned from the book.*

# TABLE OF CONTENTS

# YOU'RE DISGUSTING, BUT IN A GOOD WAY

The human body, an amazing container of life, is a wonderful world of things foul and disgusting.

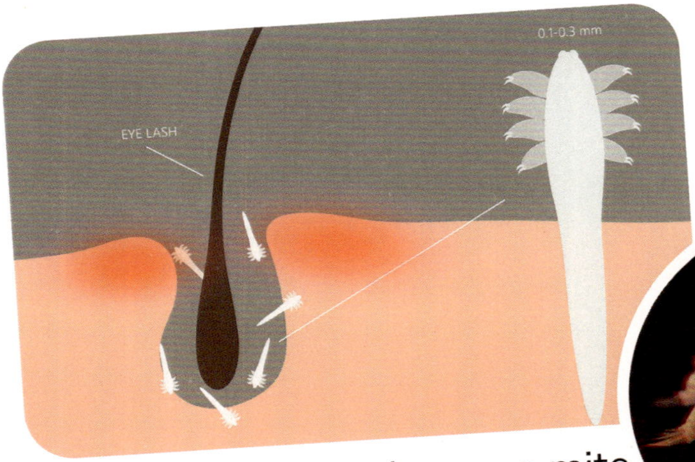

Demodex folliculorum, a mite

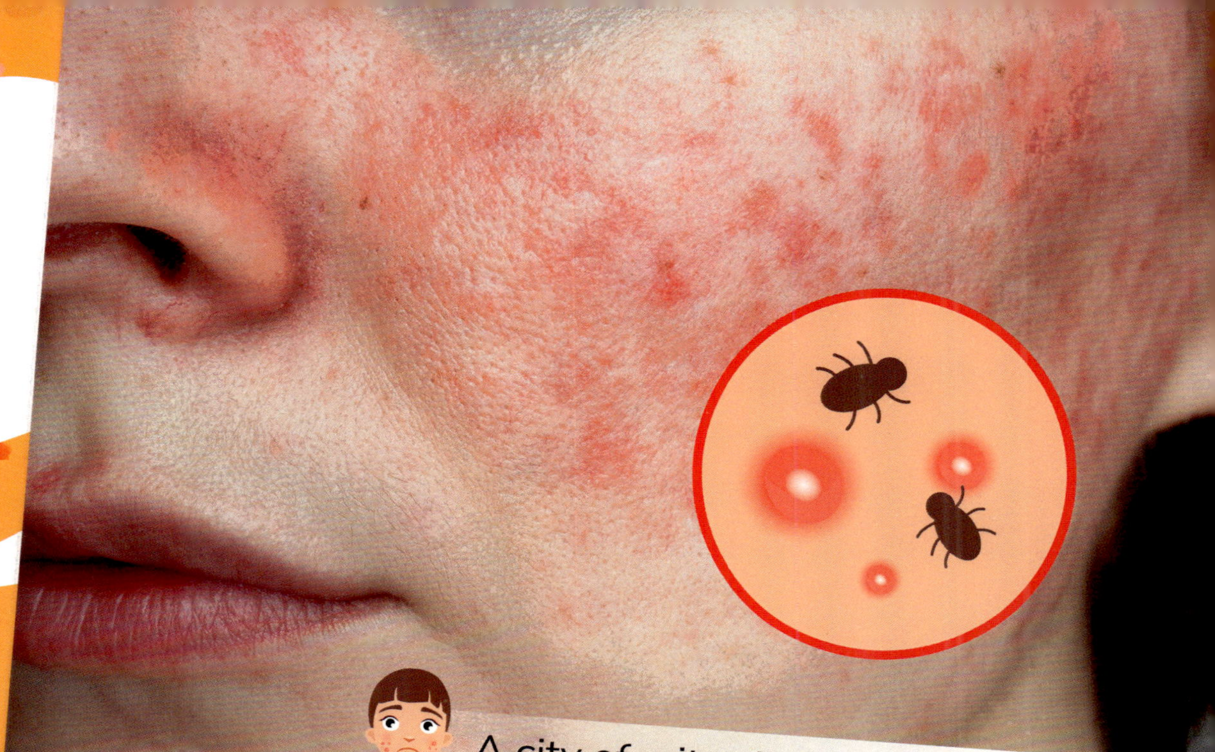

A city of mites lives on your face, quietly feasting on dead skin cells.

Ever change an overflowing diaper? For such little people, babies can really rip.

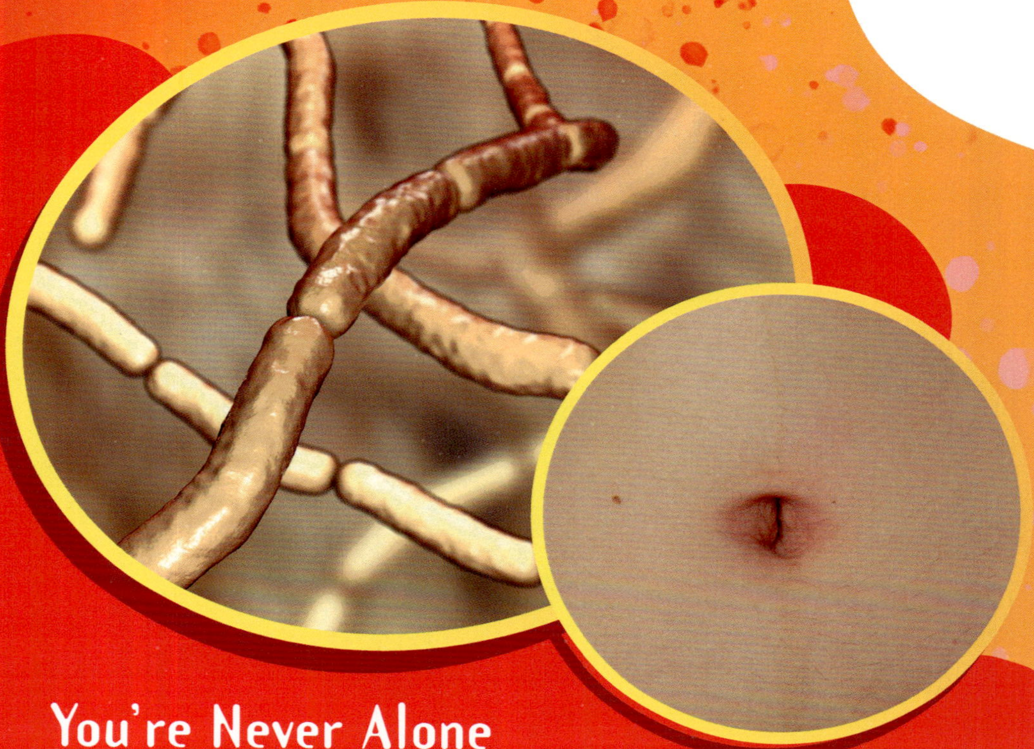

## You're Never Alone

Our bodies host **bacteria**, fungus, mites, and other life-forms unseen without a microscope. Over 60 kinds of bacteria live in our bellybuttons!

Name a body part and we can find something gross about it. From head to toe, a horror show is playing!

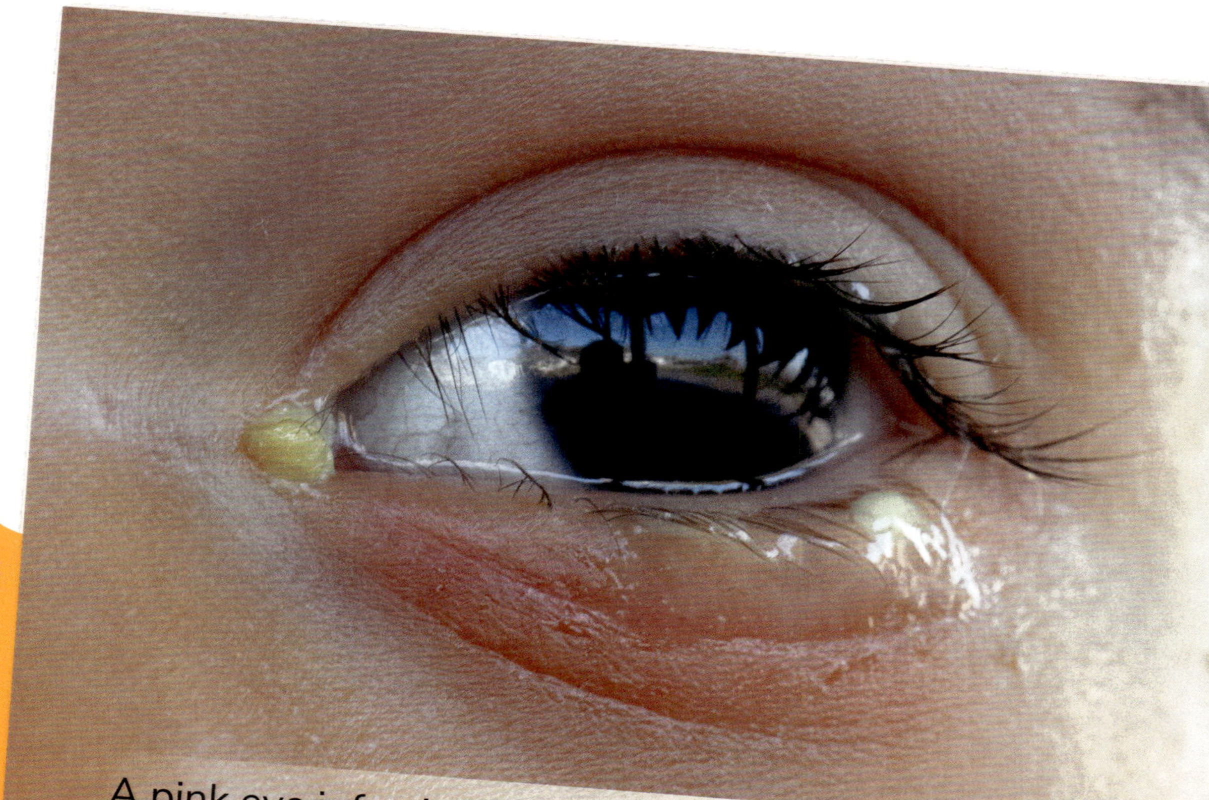

A pink eye infection makes eyes red, oozy, and crusty.

# SNOT, EAR WAX, SPIT, AND MORE

Our bodies make gross stuff. Snot and boogers result as our **mucus** cleans out dirt, dead cells, **pollen**, and other unwanted bits from our lungs and airways.

Wipe away bats in the cave with a tissue!

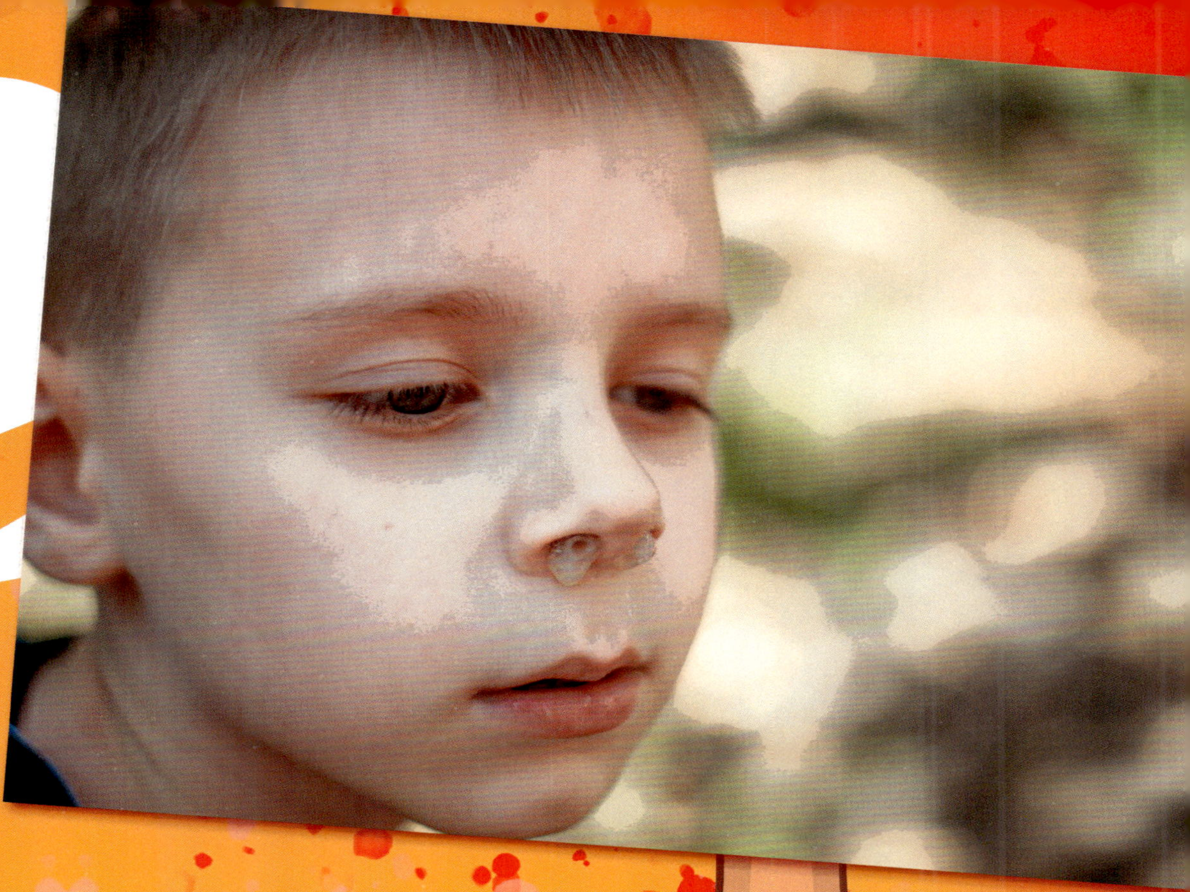

**Mighty Mucus!**
Mucus helps kill bacteria, too. Spit and snot have superpowers!

Do not fear ear wax. Like mucus, it helps get rid of dirt and dead cells. Bacteria love moist, dark places like ears. Ear wax keeps our ears healthy.

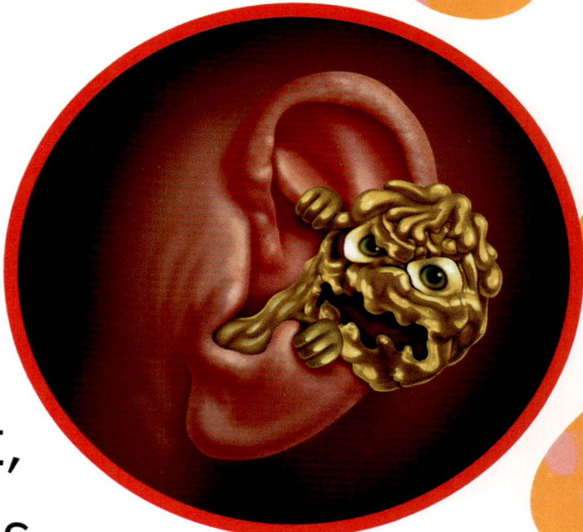

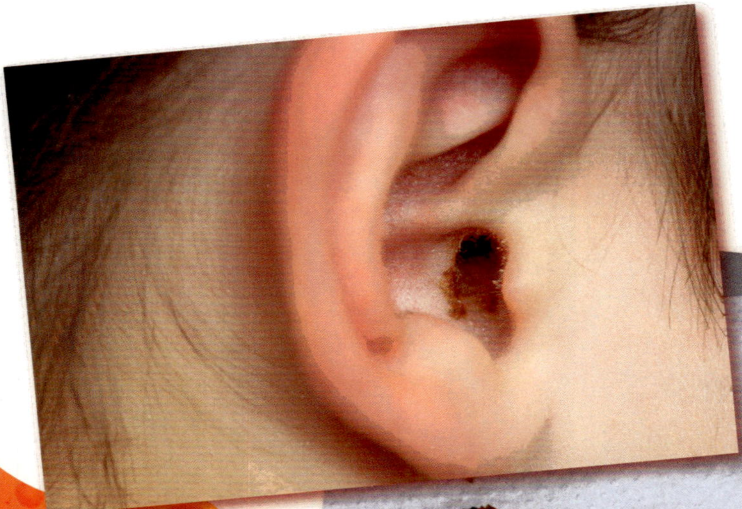

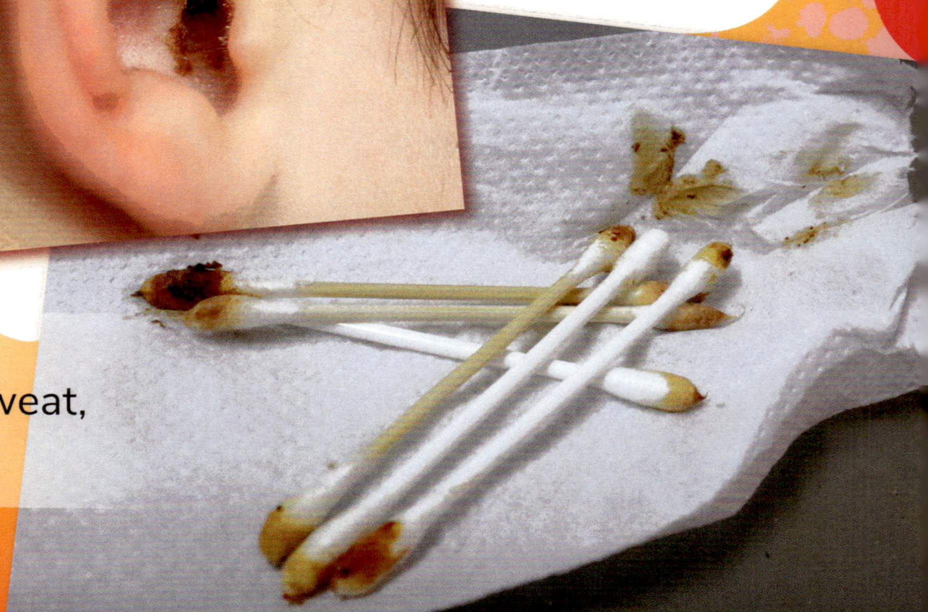

Ear wax contains oil, sweat, and dead skin.

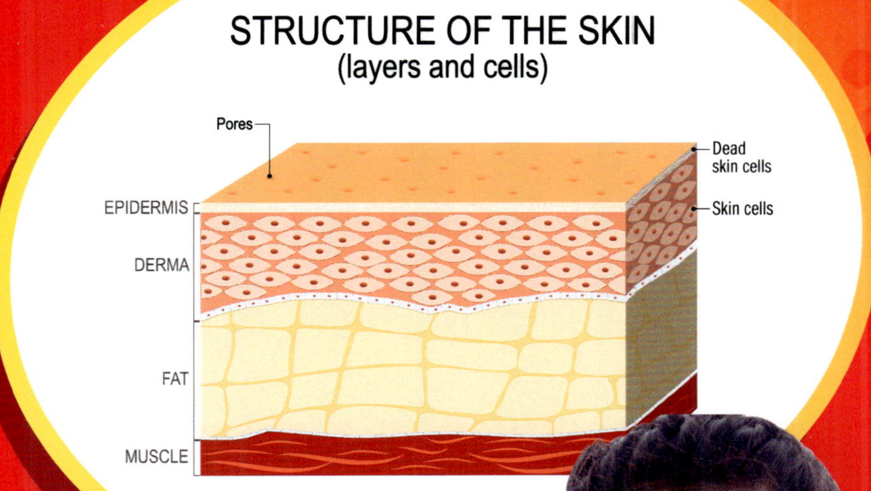


## A Flaky Fluffy?

We shed thousands of dead skin cells a minute. An adult can shed up to 9 pounds (4 kg) a year, the same weight as a small cat.

Do you need to wipe away eye gunk when you wake up? Dried tears and dust make eye boogers as our eyes clean house while we sleep.

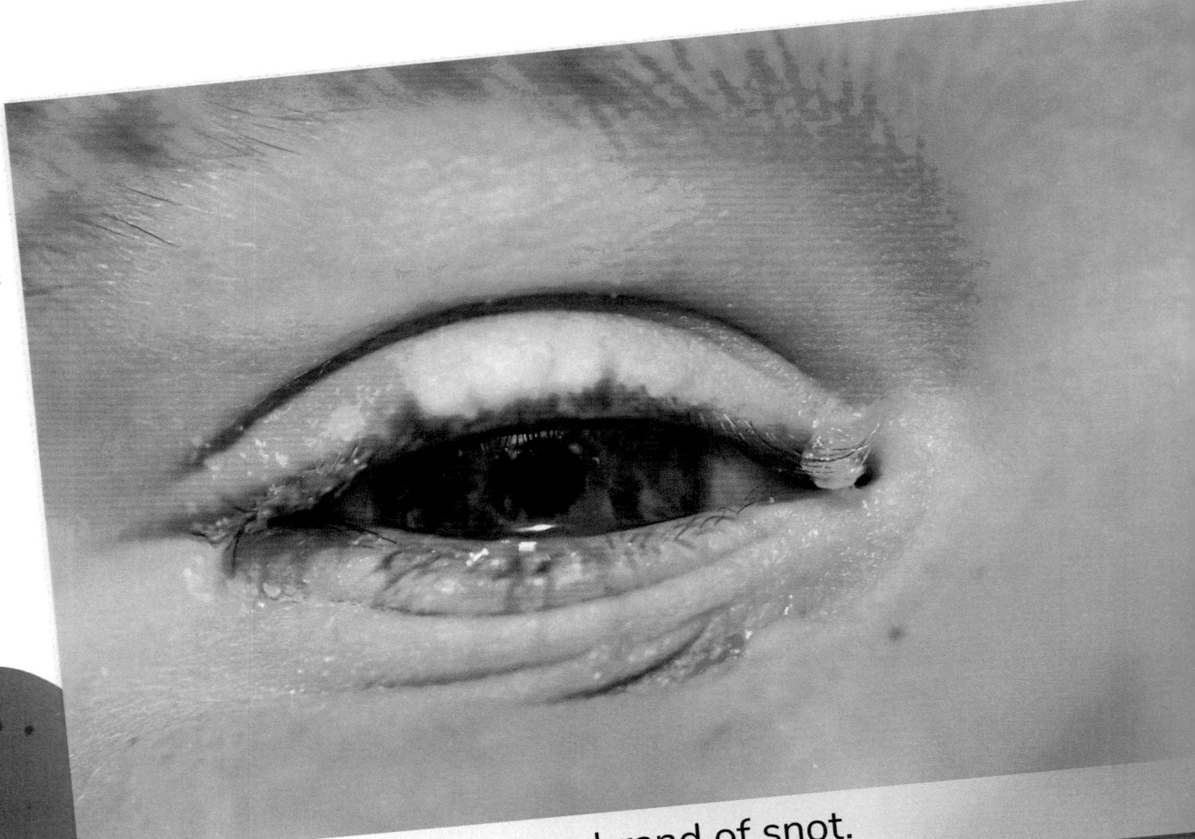

Eyes, too, make their own brand of snot.

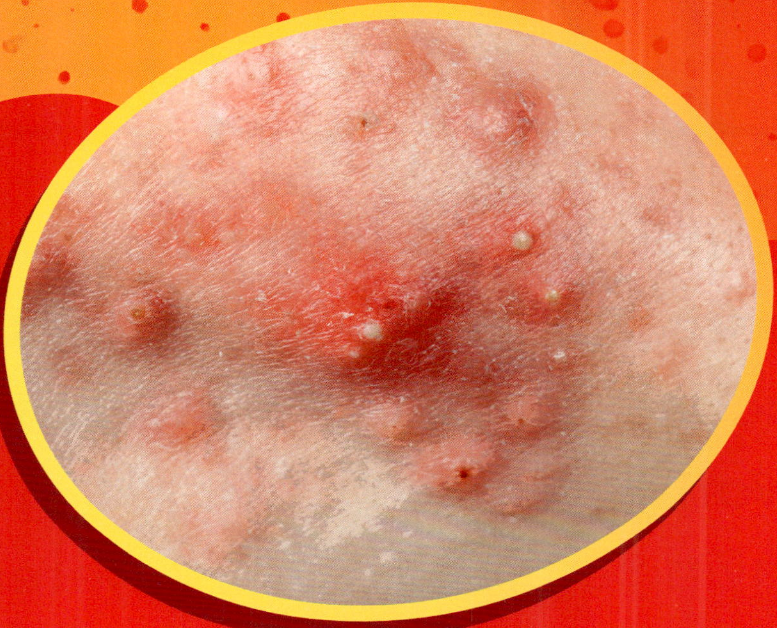

## It's the Zits

Everyone gets zits, especially teens and young adults. These little pockets of **pus** result from a minor skin infection.

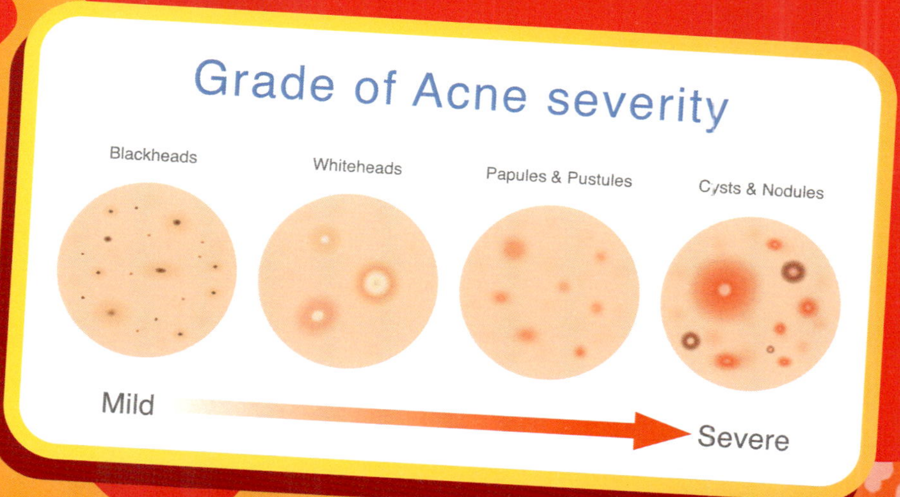

# BODY INVADERS

Sometimes harmful bacteria, parasites, and other critters attack our bodies. Head lice nestle next to our warm, cozy scalps, drinking blood and laying eggs. If only we weren't so tasty!

On close inspection, lice look like aliens.

If you get lice, don't freak out. Grab the special shampoo and a comb to kick out these unwelcome guests.

Warts love Planet Human, too! Viruses cause warts, not frogs. They enter our skin through a scratch and form a growth of extra skin cells.

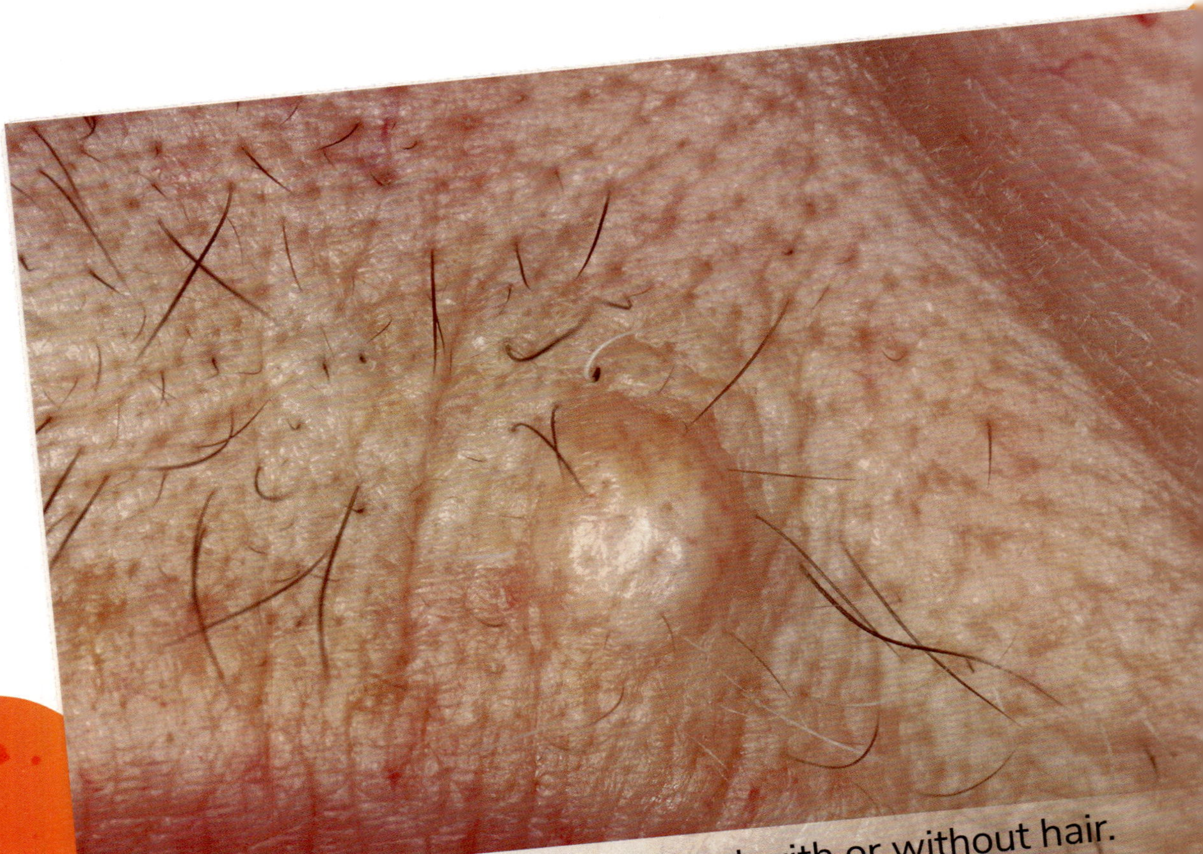

Warts can be smooth or lumpy, and with or without hair.

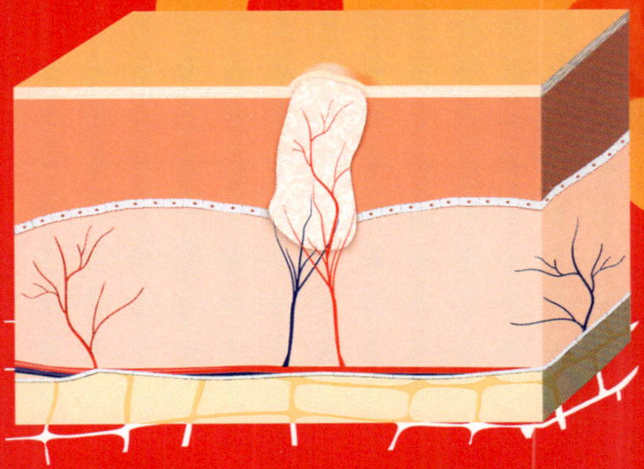

Warts commonly grow on hands and feet, but a shaving cut can lead to a face or leg wart.

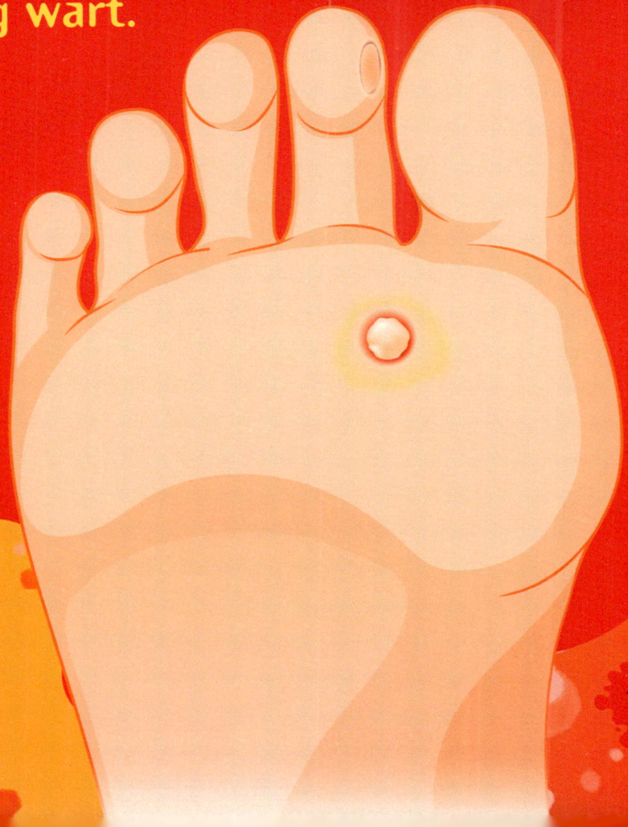

Fungus in locker rooms, showers, and around swimming pools can cause athlete's foot. This itchy infection looks like a red, scaly rash found mostly on feet and between toes.

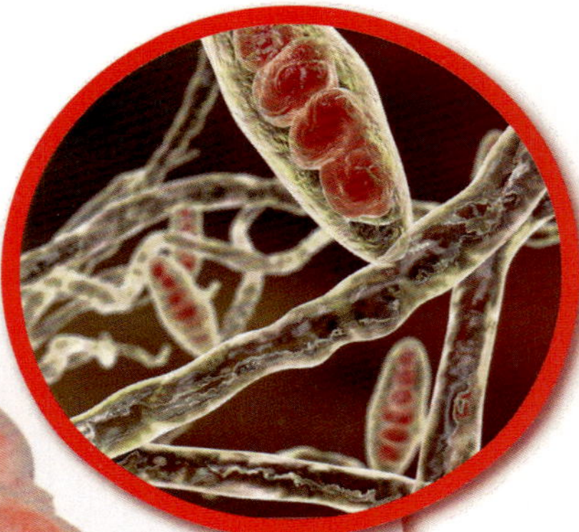

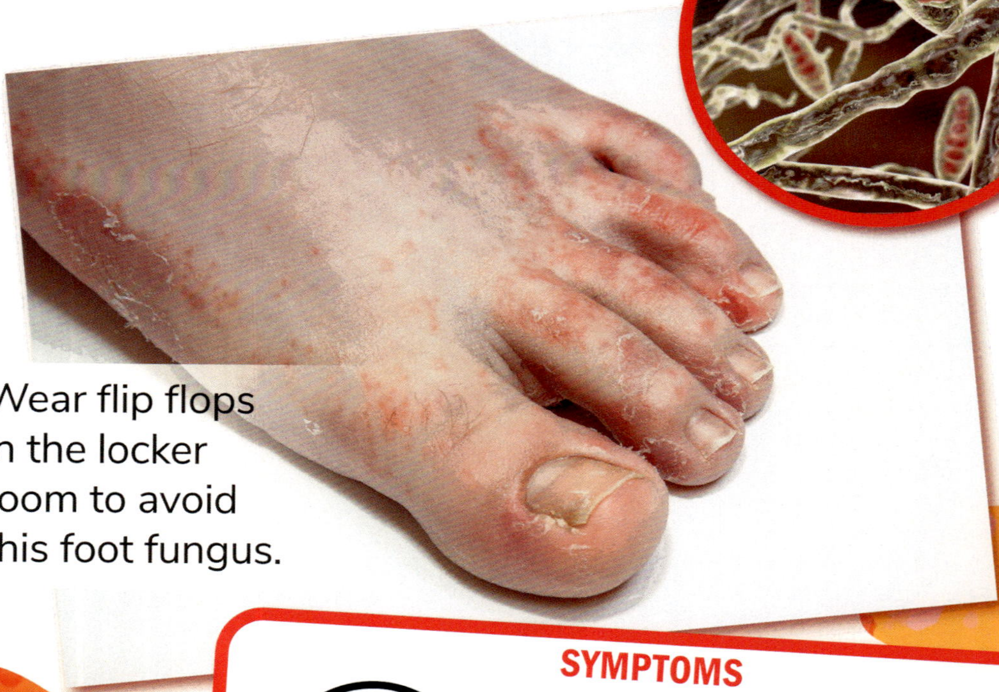

Wear flip flops in the locker room to avoid this foot fungus.

**SYMPTOMS**

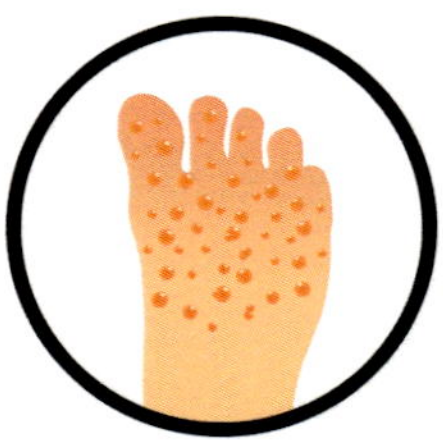

Small blisters

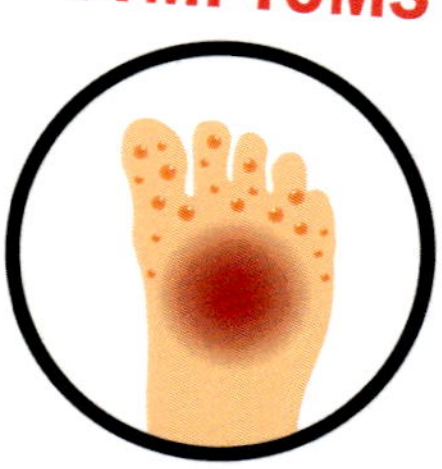

Itching with swelling and inflammation

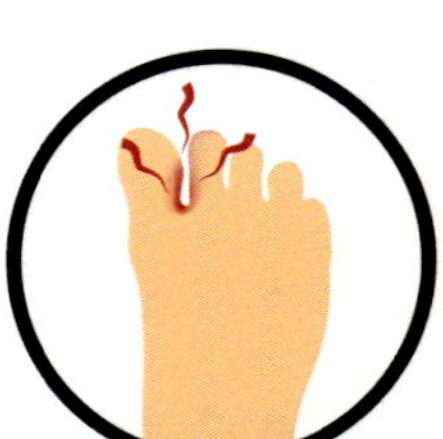

Itching and stinging between the toes

When this fungus attacks other parts of the body, we call it ringworm.

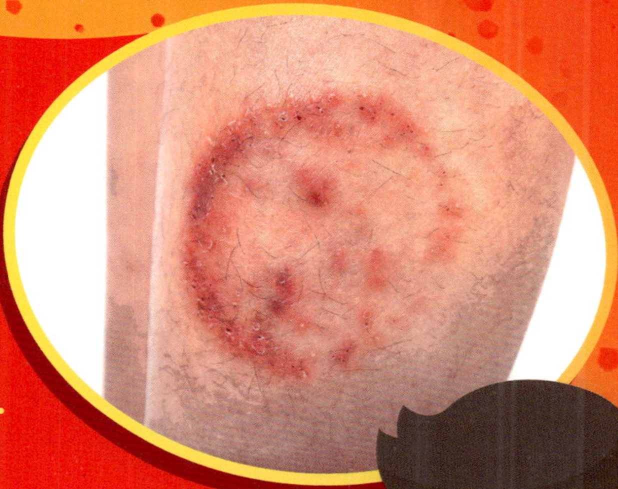

Avoid

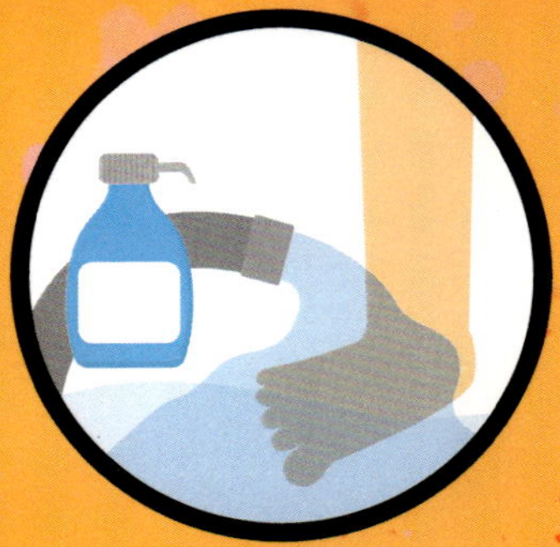

Wash your feet

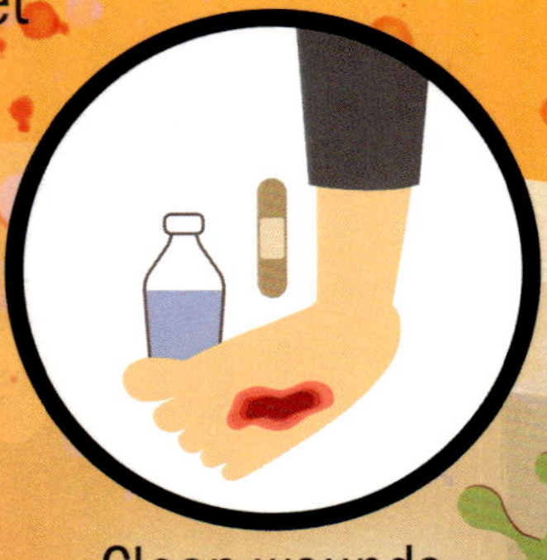

Clean wounds

When we eat food containing harmful bacteria, we get sick. Food that sits out too long, old leftovers, or contaminated fruits and vegetables can cause **diarrhea** and vomiting for hours.

Diarrhea and puke help rid the body of harmful bacteria that cause food poisoning.

## The Winter Vomiting Bug

Norovirus spreads fast through food and drink and other sick people. Diarrhea, vomiting, and painful stomach cramps make life miserable for a few days.

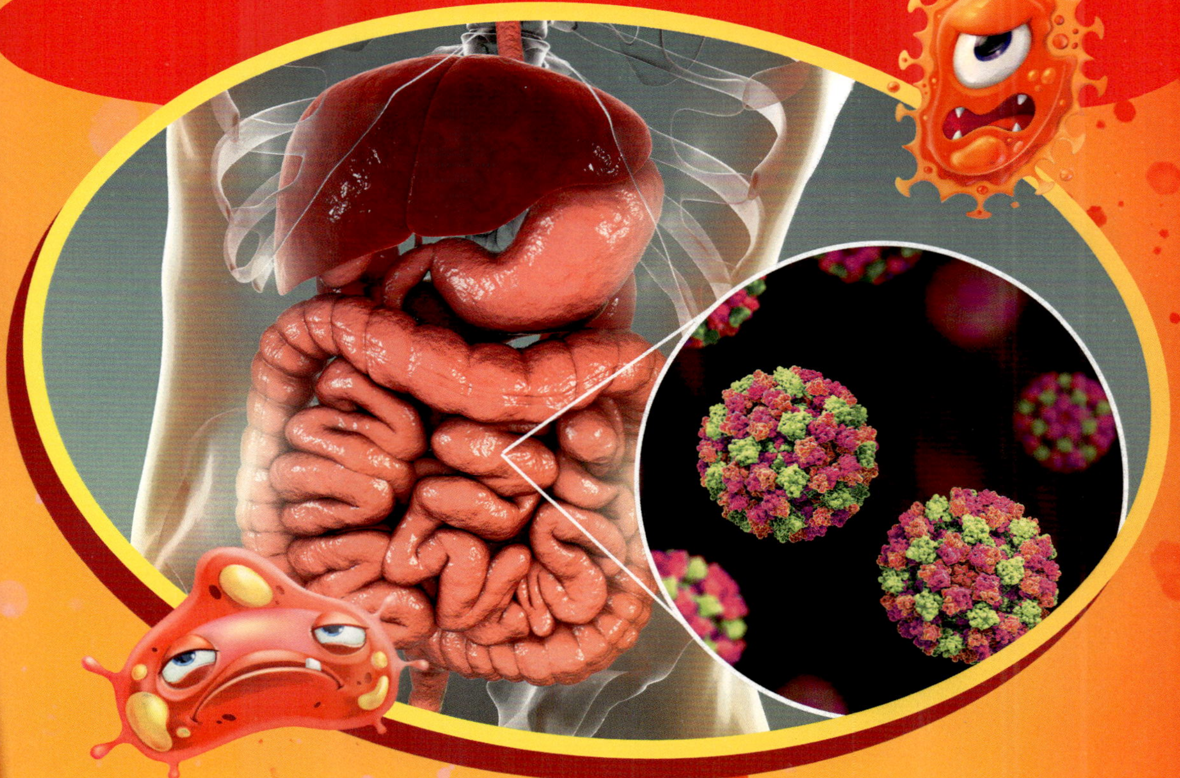

### Nicknames

Diarrhea has lots of other names. Have you heard these?

- the trots
- the squirts
- meals on wheels
- toilet chili
- wet dragon

# MORE BODY BATTLES

Our bodies constantly defend against disease. In our mouths, our spit battles bacteria that causes tooth decay and the yeast that causes **thrush.**

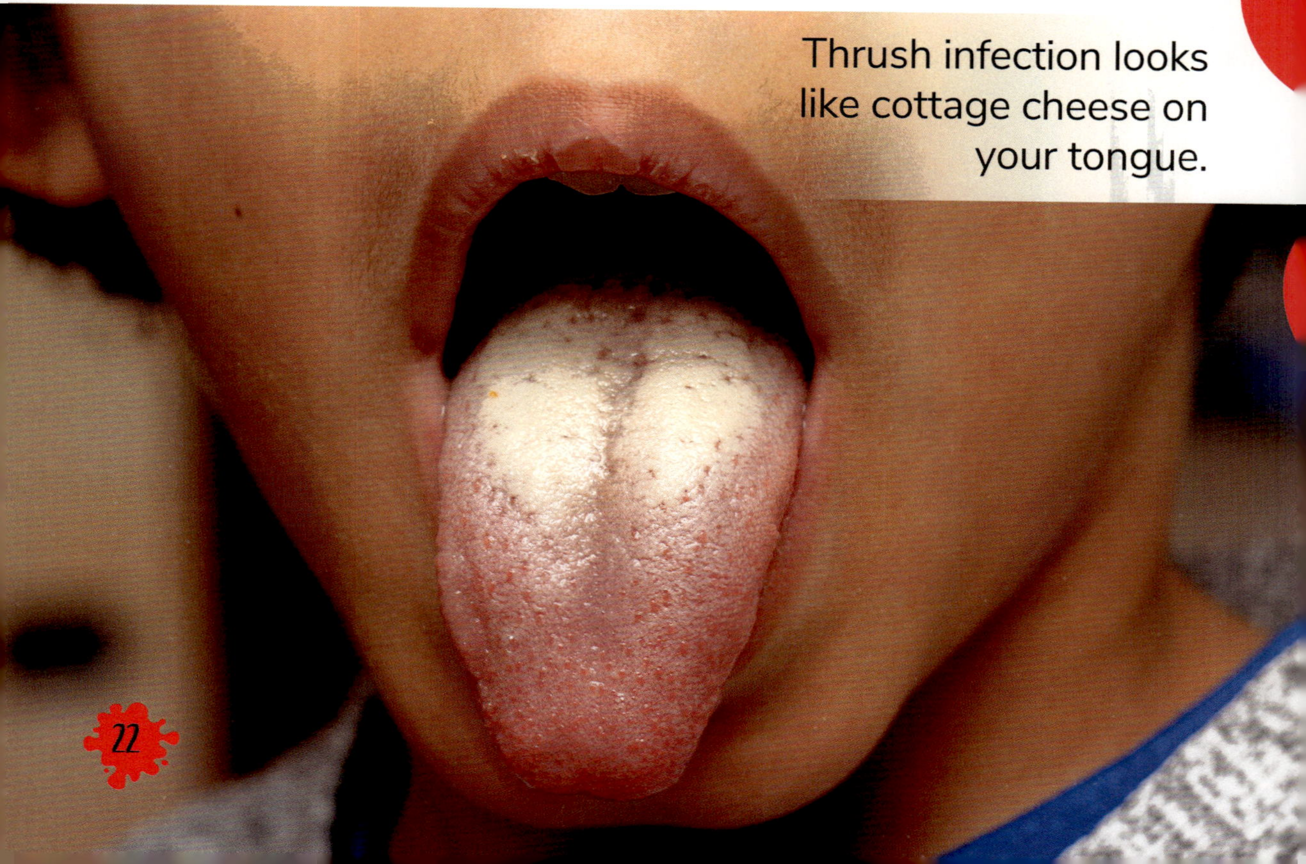

Thrush infection looks like cottage cheese on your tongue.

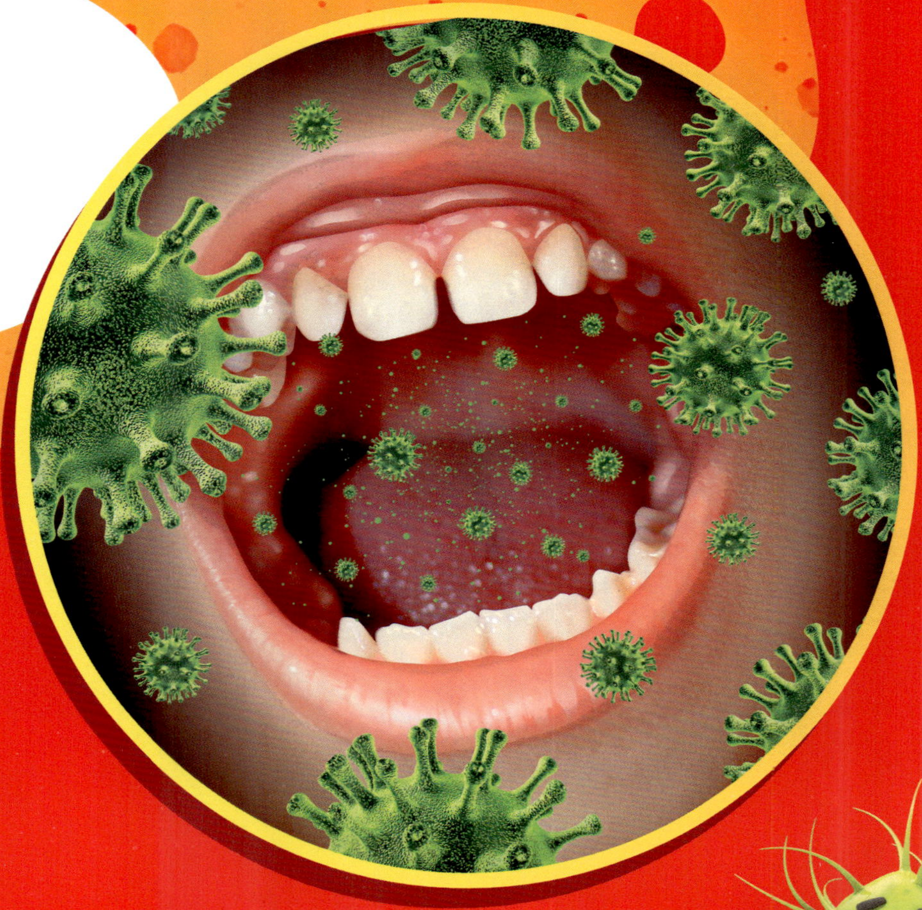

**No Fuzzy Teeth!**
Bacteria and food leave **plaque** deposits on our teeth. Plaque eats away the hard surface, leading to cavities.

Colds can lead to painful ear infections. Infected **fluid** builds behind your ear drum, causing pain and muffling sound.

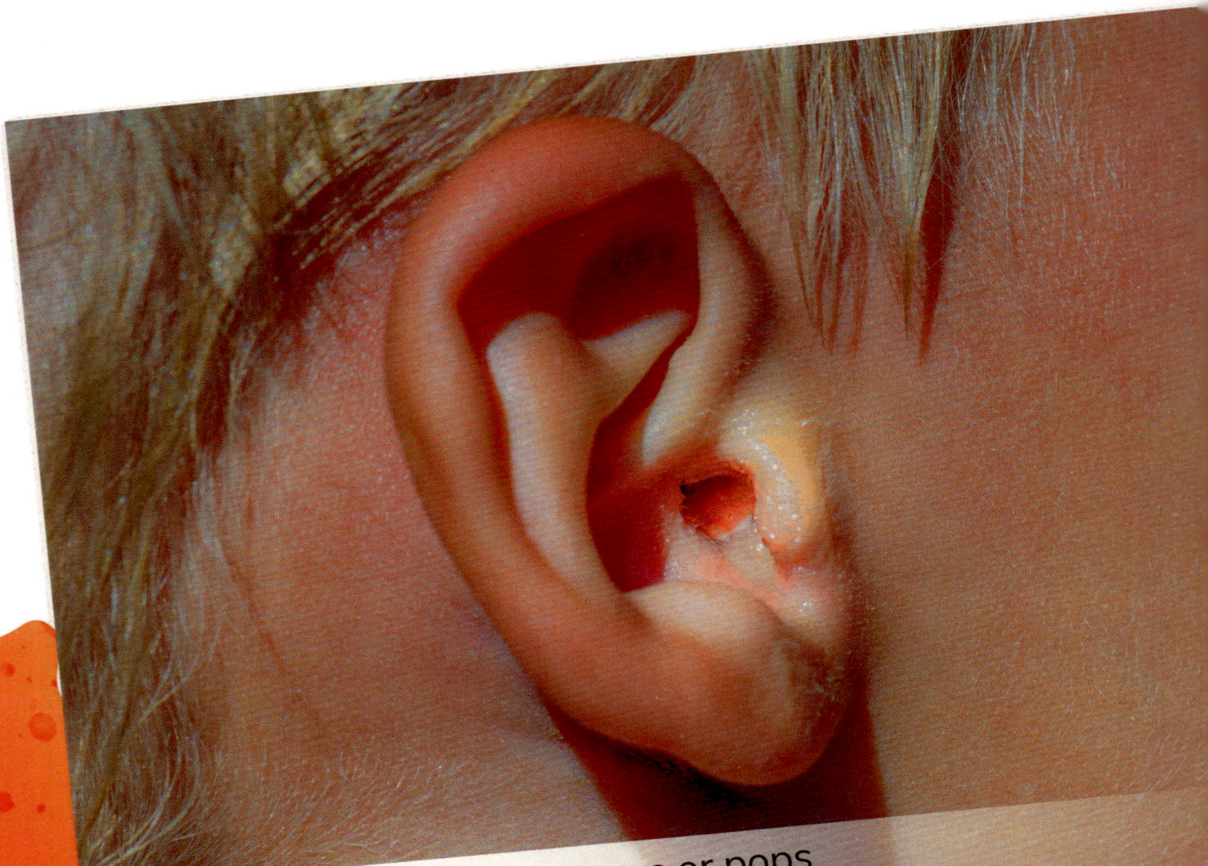

Sometimes, the ear drum tears or pops and incredibly stinky pus leaks out.

# OUTER EAR INFECTION

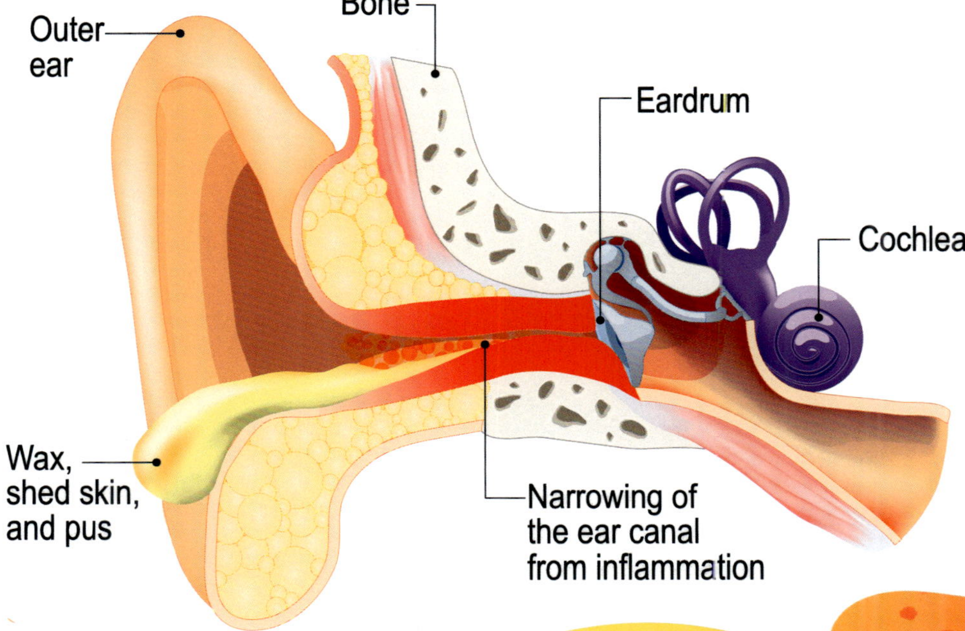


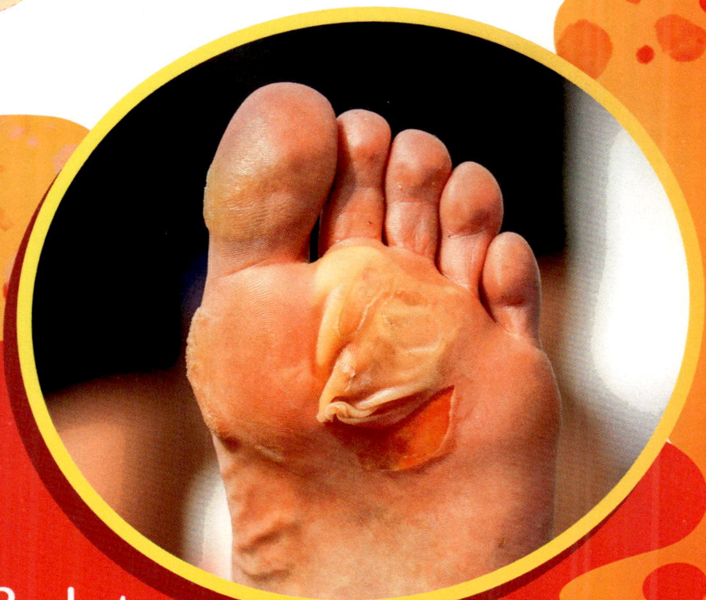

**Water Pockets**

Blisters can form from injuries such as burns, or when tight shoes rub our skin.

**Wounds** and other injuries can also get infected. Pus and fluids drain out. Tumors, too, can grow and spread. Some are harmless, while others need treatment to stop their advance.

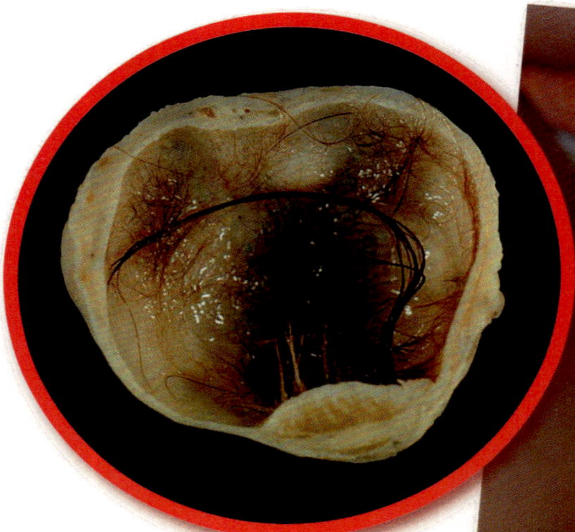

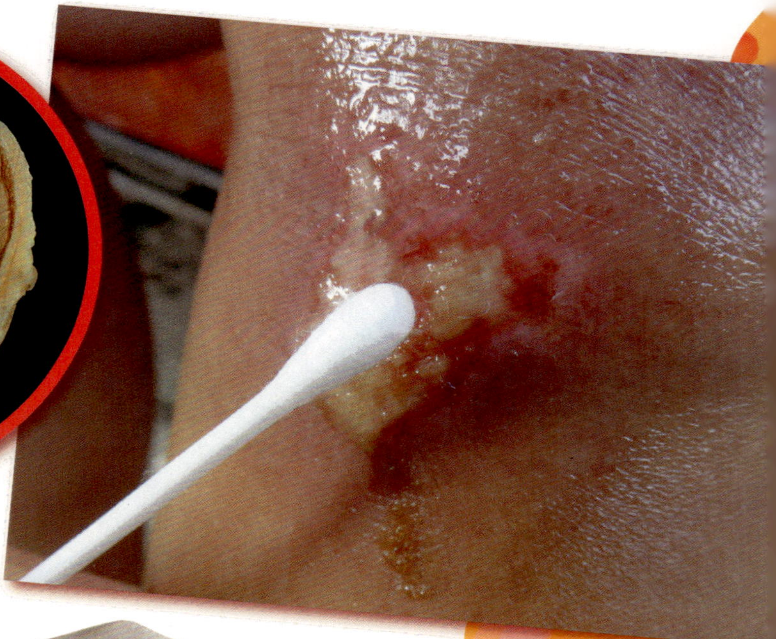

Some tumors can grow teeth and hair!

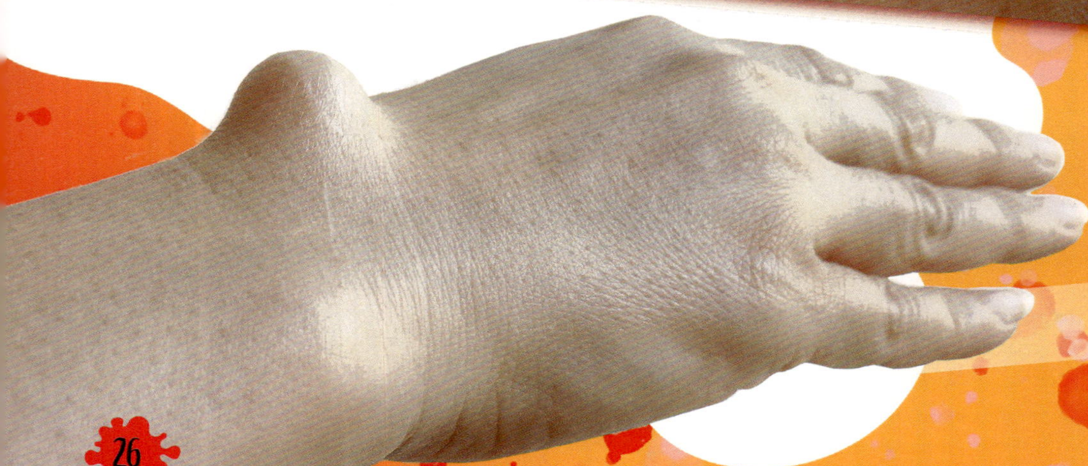

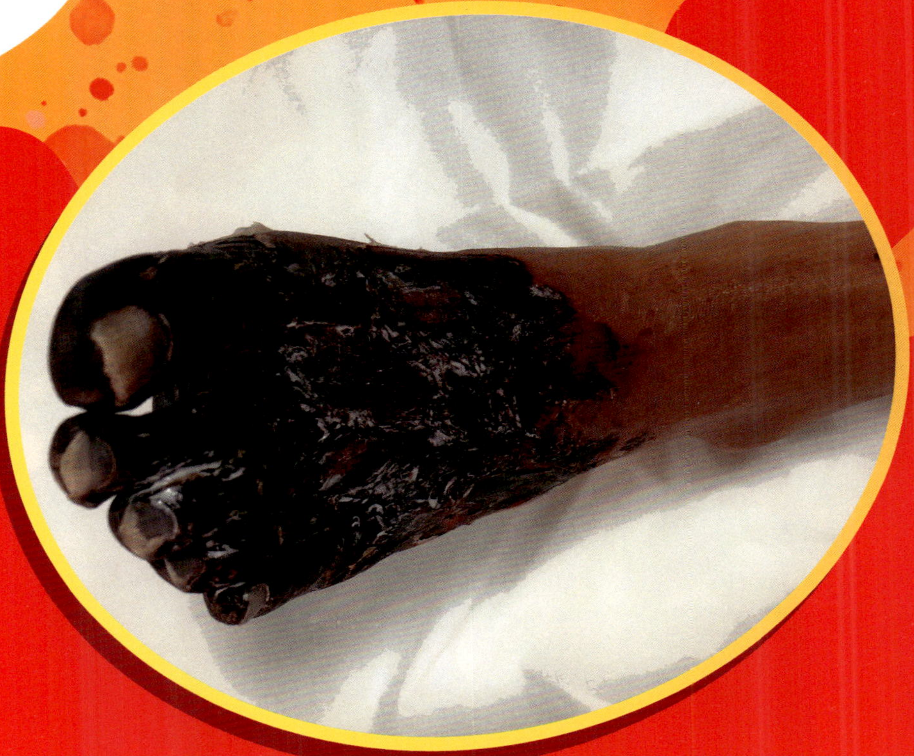

## Extremely Gross

Poor blood flow to damaged and infected areas can lead to **gangrene**. Toes especially may turn black and die. If not stopped, gangrene can cause death as the infection spreads.

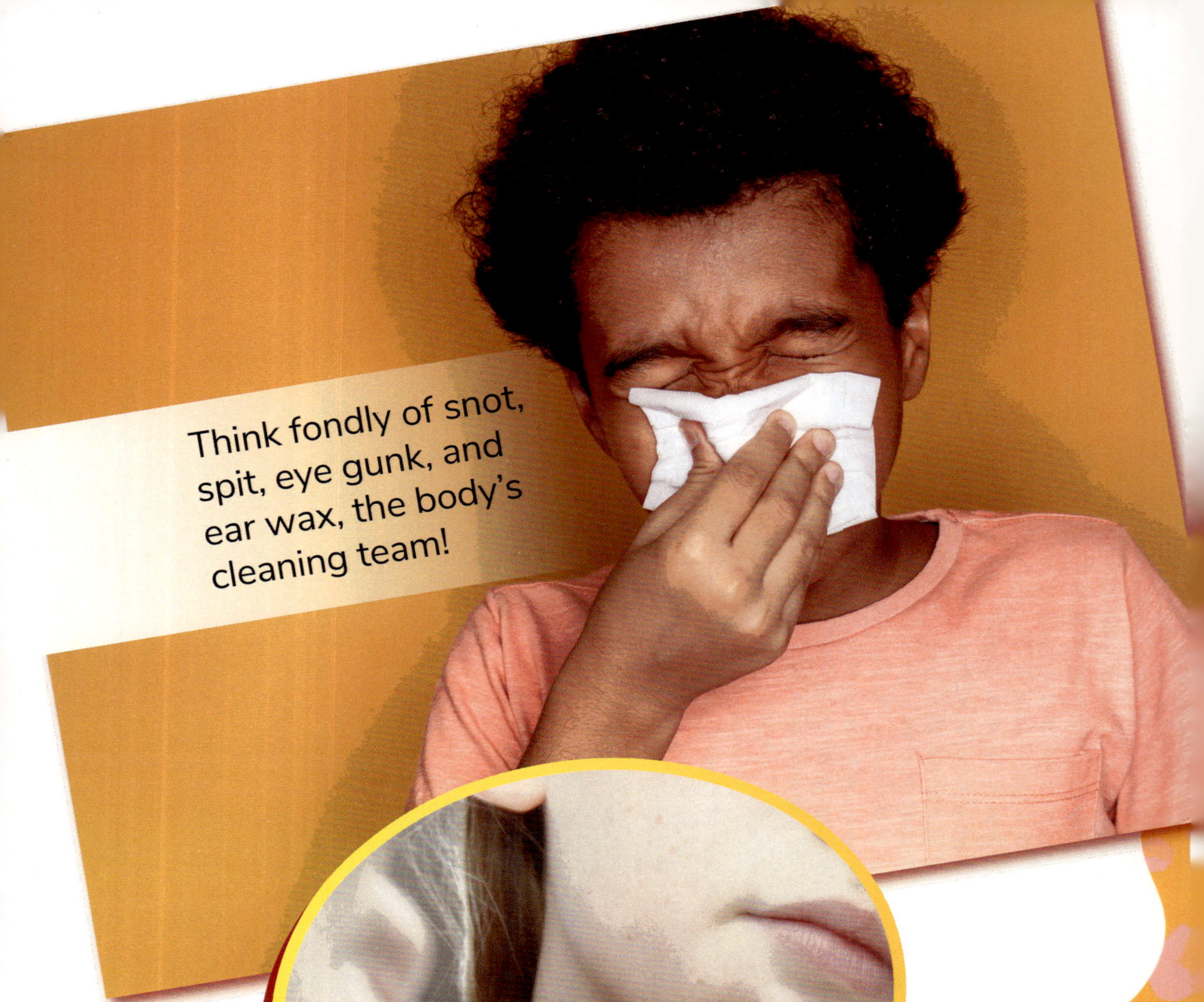

Think fondly of snot, spit, eye gunk, and ear wax, the body's cleaning team!

## Let the Body Work

Scabs help scrapes heal. Picking can lead to infection or scars.

Treat your body well. Every day it battles to keep you clean, healthy, and at your best.

# GLOSSARY

**bacteria** (back-TEER-ee-uh): Tiny living things that break down food, and may cause illness

**diarrhea** (die-uh-REE-uh): Loose, watery, or slimy poop

**fluid** (FLEW-id): A liquid or substance that flows easily

**gangrene** (gang-GREEN): Death of body parts or tissues, caused by poor blood flow or infection

**mucus** (MEW-kiss): A slimy fluid made by body tissues from the lungs to the nose for cleaning and defense

**plaque** (PLACK): Soft or hard mixture of food, spit, and bacteria that forms on teeth and causes decay

**pollen** (PAHL-en): Fine grains released into the air by plants as part of reproduction

**pus** (PUHS): Fluid made by the body in response to infected injuries

**thrush** (THRUSH): A common mouth infection caused by a certain kind of yeast

**wounds** (WOONDZ): Injuries that are open, such as a cut or scrape

# INDEX



# WEBSITES TO VISIT

www.ducksters.com/science/bacteria.php

https://kidshealth.org/en/kids/center/htbw-main-page.html

www.konnecthq.com/human-body/

# ABOUT THE AUTHOR

## Julie K. Lundgren

Julie K. Lundgren grew up on the north shore of Lake Superior, a place of woods, water, and adventure. She loves bees, dragonflies, old trees, and science and has a special place in her heart for disgustingly cool animals. Her interests led her to a degree in biology and a lifelong curiosity about wild places.

**Crabtree Publishing**

crabtreebooks.com 800-387-7650



Produced by: Blue Door Education for Crabtree Publishing
Written by: Julie K. Lundgren
Designed by: Jennifer Dydyk
Edited by: Tracy Nelson Maurer
Proofreader: Crystal Sikkens

| | |
|---|---|
| Hardcover | 978-1-4271-5451-4 |
| Paperback | 978-1-4271-5457-6 |
| Ebook (pdf) | 978-1-4271-5463-7 |
| Epub | 978-1-4271-5469-9 |
| Read-along | 978-1-4271-5475-0 |
| Audio book | 978-1-4271-5481-1 |

Printed in Canada/022024/CPC20240130

**Published in Canada**
**Crabtree Publishing**
616 Welland Avenue
St. Catharines, Ontario
L2M 5V6

**Published in the United States**
**Crabtree Publishing**
347 Fifth Avenue
Suite 1402-145
New York, NY 10016

Cover photo © sumroeng chinnapan/ Shutterstock.com, cover splat art on cover and throughout © SpicyTruffel pg. 4 illustration © visona29 photo © Kalcutta pg. 5 top photo © Yuliya Shauerman illustration in circle and of girl © Igdeeva Alena bottom photo © vchal baby illustration © Adrian Isaia pg. 6 large photo © Kateryna Kon smaller photo © Bokeh Blur Background pg. 7 © sruilk pg. 8 © diy13 pg. 9 photo © Zapylaieva Hanna illustration © Top Vector Studio pg. 10 (top) © Lightspring (center) © Andrii Spy_k Richokphoto pg. 11 photo © all_about_people illustration © Designua pg. 12 © ElRoi pg. 13 (top) © Bencemor (bottom) © solar22 pg. 14 lice © Protasov AN magnifying glass © Romolo Tavani pg. 15 photo © Lapina lice © Refluo boy illustrations © bego pg. 16 © Henrik Larsson pg. 17 © Designua pg.s 18-19 bacteria © Kateryna Kon foot photo © Miroslav Lukic illustrations © Love You Stock ringworm photo © Photo Win1 pg.s 20-21 bacteria cartoons © Natykach Nataliia girl © Daisy Daisy intestines © Kateryna Kon pg. 22 © sruilk pg. 23 (top) © Lightspring (bottom) © Giovanni Cancemi pg. 24 © weakiva pg. 25 (top) © Designua (bottom) © Ron Hoenson pg. 26 tumor with hair public domain image by Ed Uthman, MD, wound © AppleDK hand © DaViDa S pg. 27 © Duangnapa Kanchanasakun pg. 28 (top) © Bernardo Emanuelle (bottom) © Cheryl Casey pg. 29 © Versta. All images from Shutterstock.com except tumor photo page 26.

**Library and Archives Canada Cataloguing in Publication**

Title: Gross and disgusting stuff in your body / Julie K. Lundgren.
Names: Lundgren, Julie K., author.
Description: Series statement: Gross and disgusting things | "A Crabtree branches book". | Includes index.
Identifiers: Canadiana (print) 20210220481 | Canadiana (ebook) 2021022049X | ISBN 9781427154514 (hardcover) | ISBN 9781427154576 (softcover) | ISBN 9781427154637 (HTML) | ISBN 9781427154699 (EPUB) | ISBN 9781427154750 (read-along ebook)
Subjects: LCSH: Human body—Juvenile literature. | LCSH: Human body—Miscellanea—Juvenile literature.
Classification: LCC QP37 .L86 2022 | DDC j612—dc23

**Library of Congress Cataloging-in-Publication Data**

Names: Lundgren, Julie K., author.
Title: Gross and disgusting stuff in your body / Julie K. Lundgren.
Description: New York, NY : Crabtree Publishing Company, [2022] | Series: Gross and disgusting things - a Crabtree branches book | Includes index.
Identifiers: LCCN 2021022341 (print) | LCCN 2021022342 (ebook) | ISBN 9781427154514 (hardcover) | ISBN 9781427154576 (paperback) | ISBN 9781427154637 (ebook) | ISBN 9781427154699 (epub) | ISBN 9781427154750
Subjects: LCSH: Human body--Juvenile literature. | Human body--Miscellanea--Juvenile literature.
Classification: LCC QP37 .L86 2022 (print) | LCC QP37 (ebook) | DDC 612--dc23
LC record available at https://lccn.loc.gov/2021022341
LC ebook record available at https://lccn.loc.gov/2021022342